Connect the Dots for Adults
Large Print Animal Kingdom

By Mindful Coloring Books

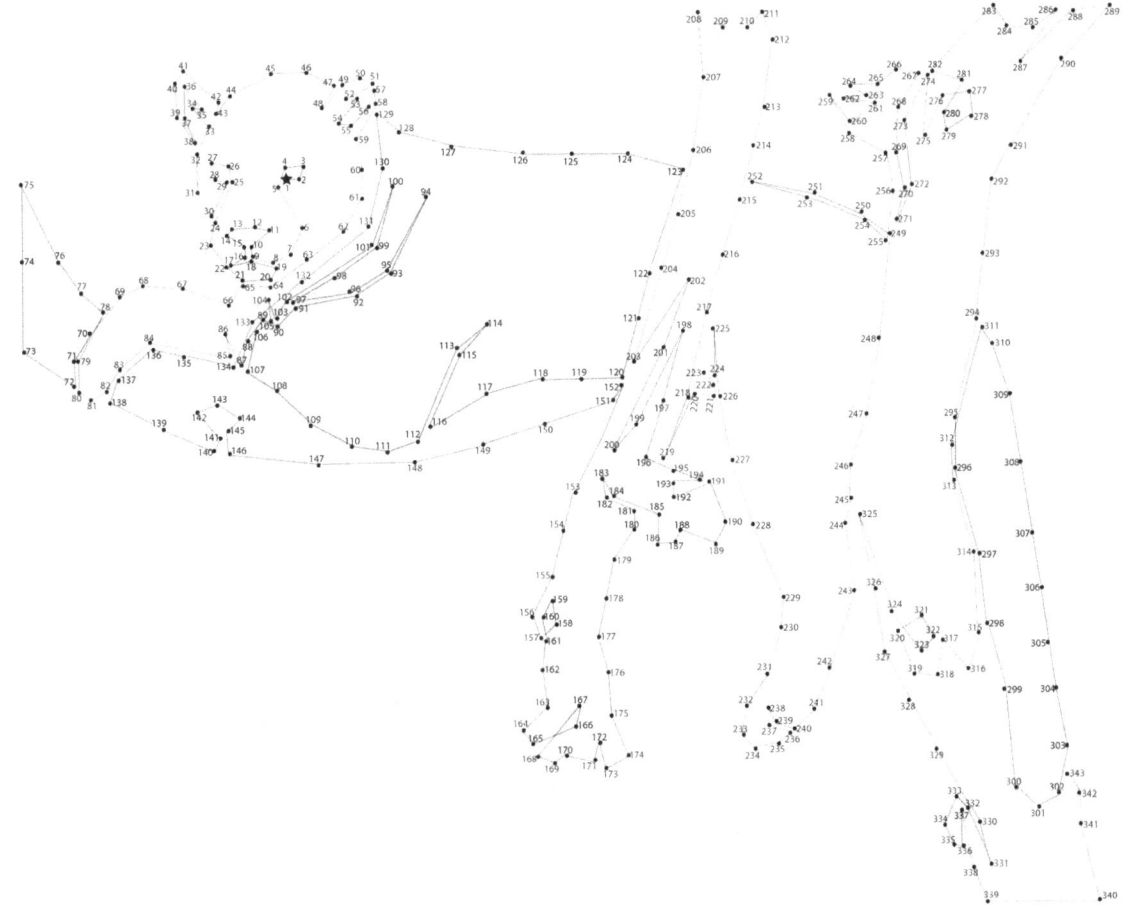

Copyright © 2018 by Mindful Coloring Books

DOT-TO-DOT PUZZLES AREN'T JUST FOR KIDS!

Connecting dots to reveal a picture can be fun
and relaxing at any age!

Directions are very simple:

Find the dot marked 1 then connect it to the
dot marked 2 and continue with the next number
until the last dot.

Don't stress, each dot is numbered and there
is no time limit. All completed puzzles are
included in the back if you need a hint.

Be sure you have fun!

A few other tips:

Want a bigger challenge? Try starting at a
number other than 1. Or try going backwards.

Use a pencil so you can erase mistakes.

Keep your utensil tip sharp for finer lines. Most
people think fine lines make a better looking
completed picture.

Finished the puzzle? Color it! Use colors and
your imagination to really make the page your own.

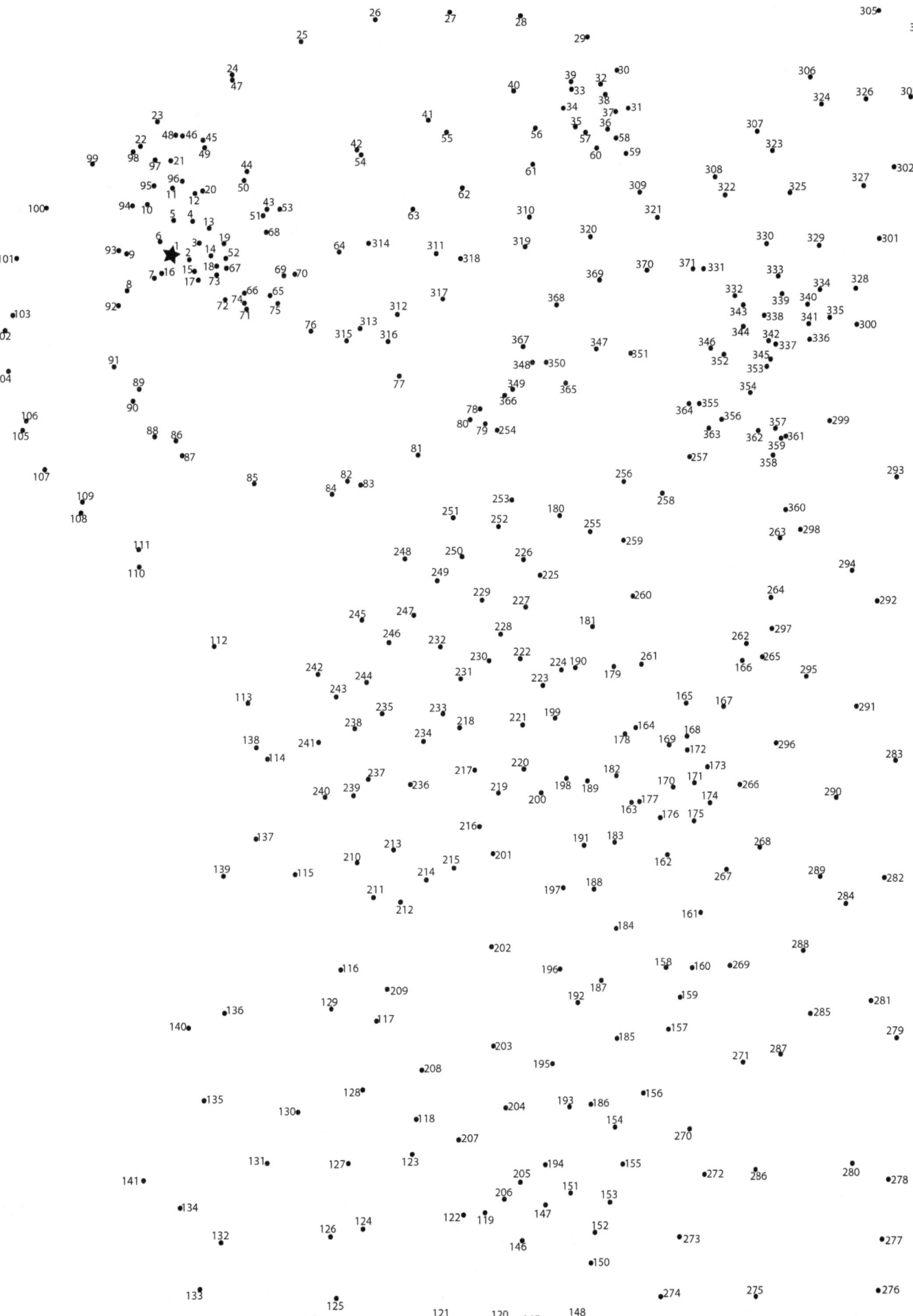

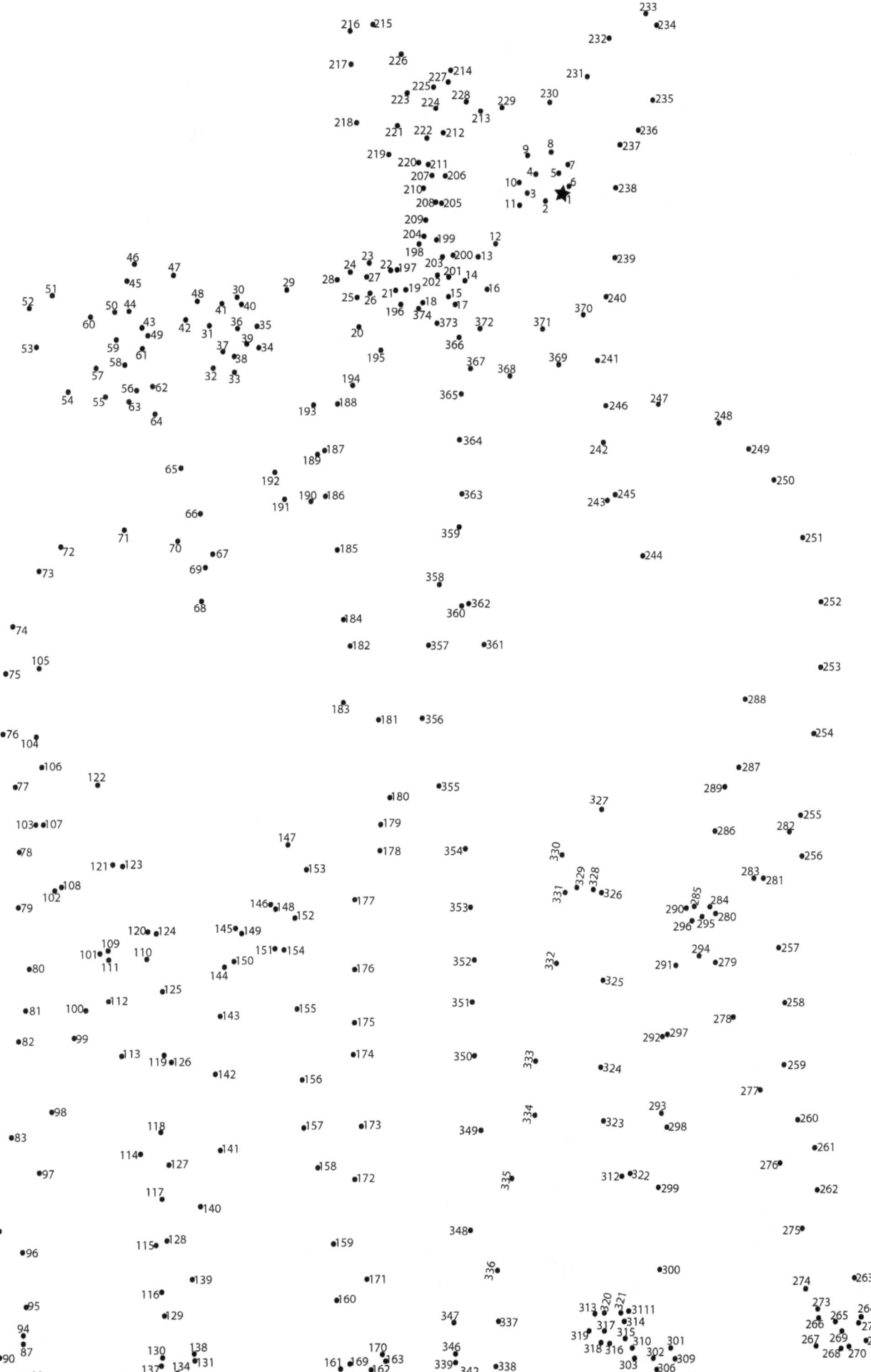

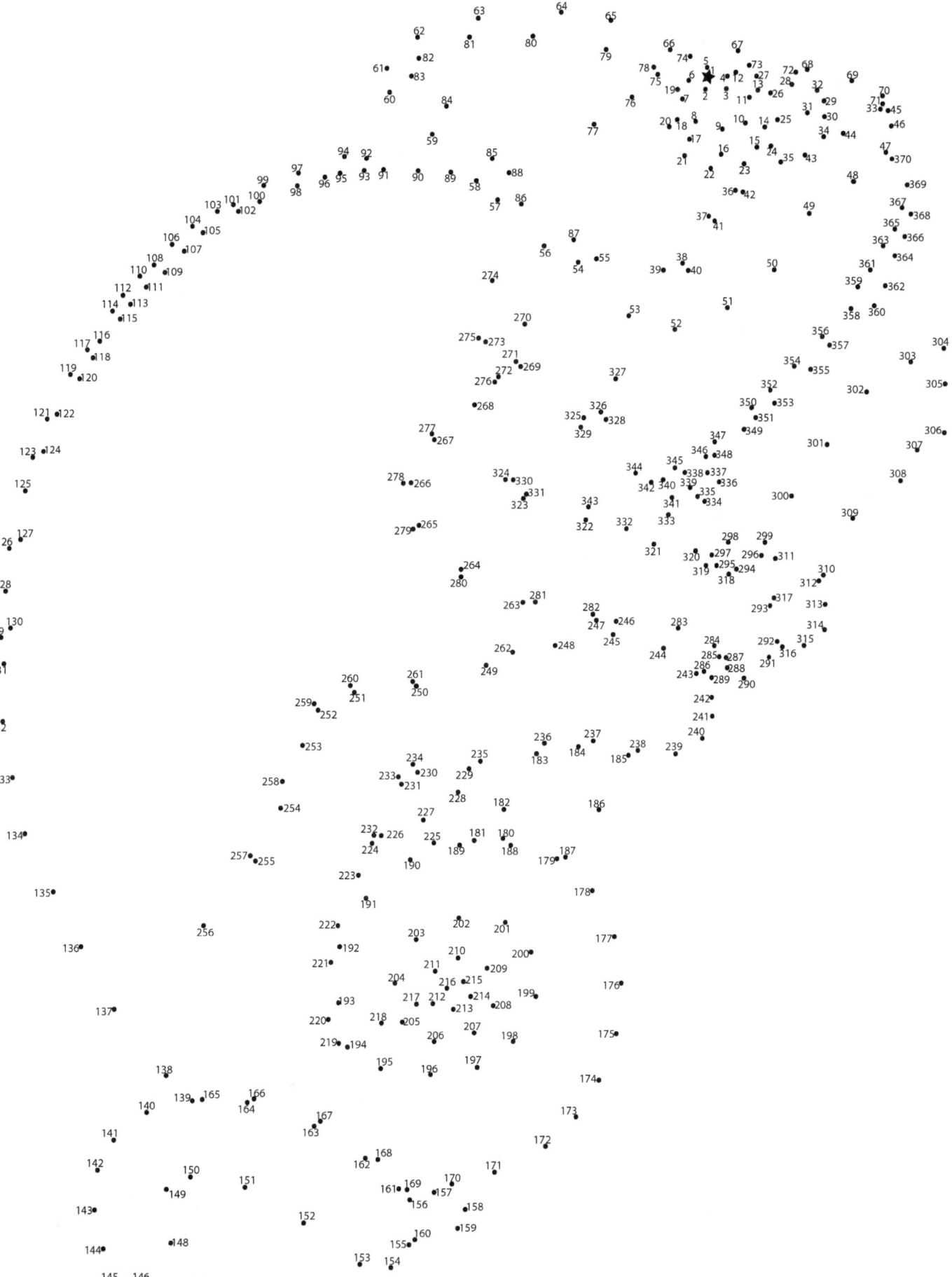

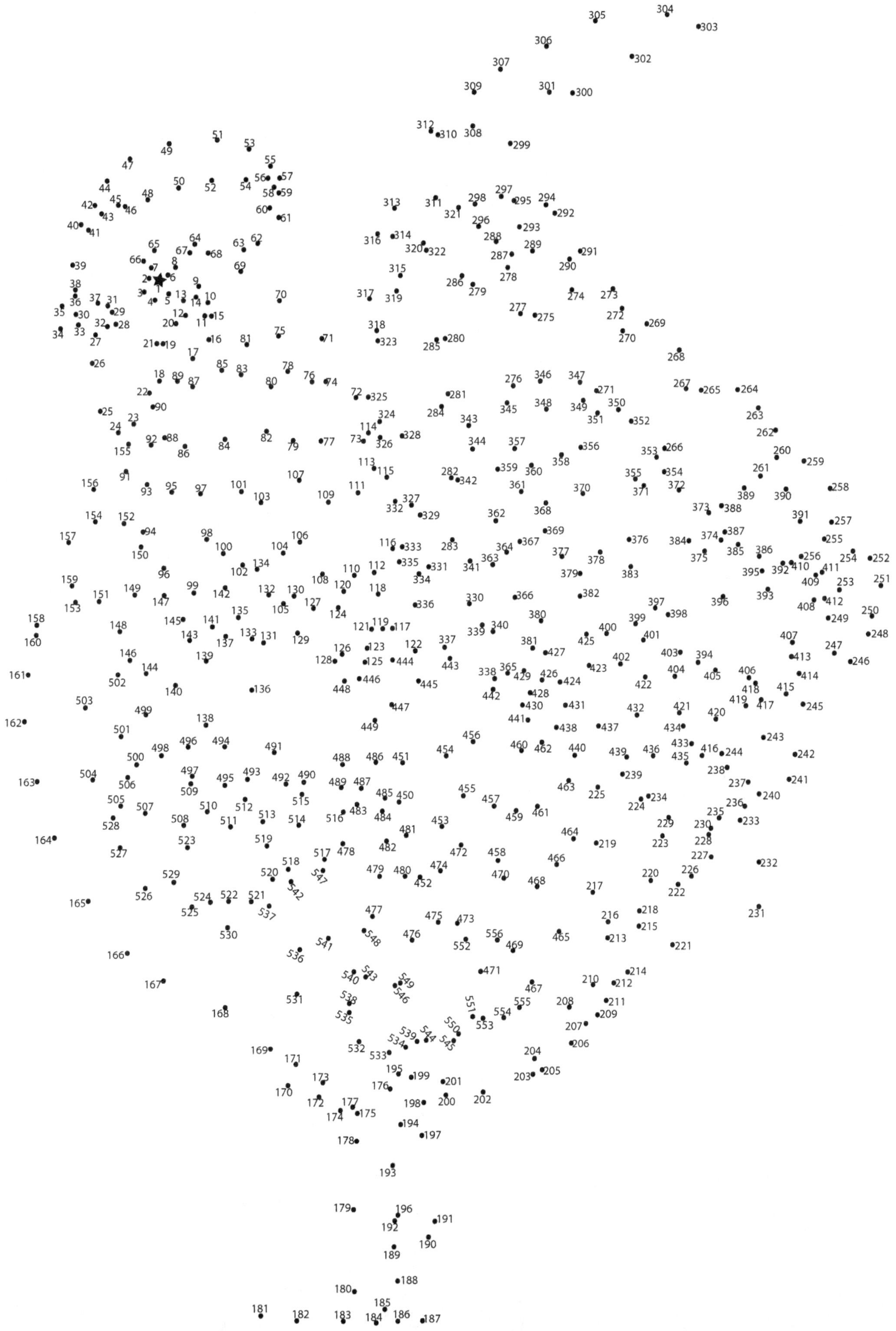

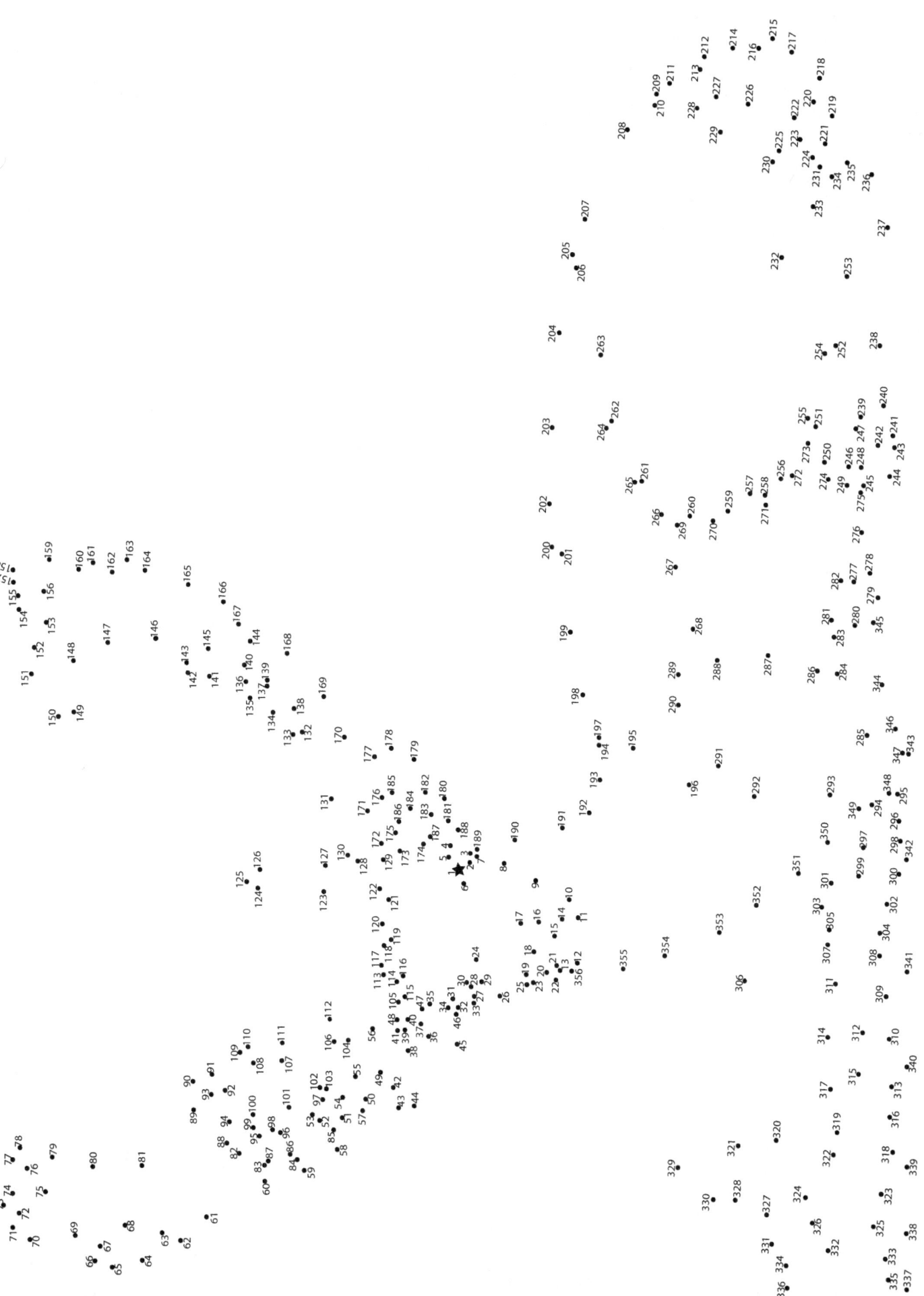

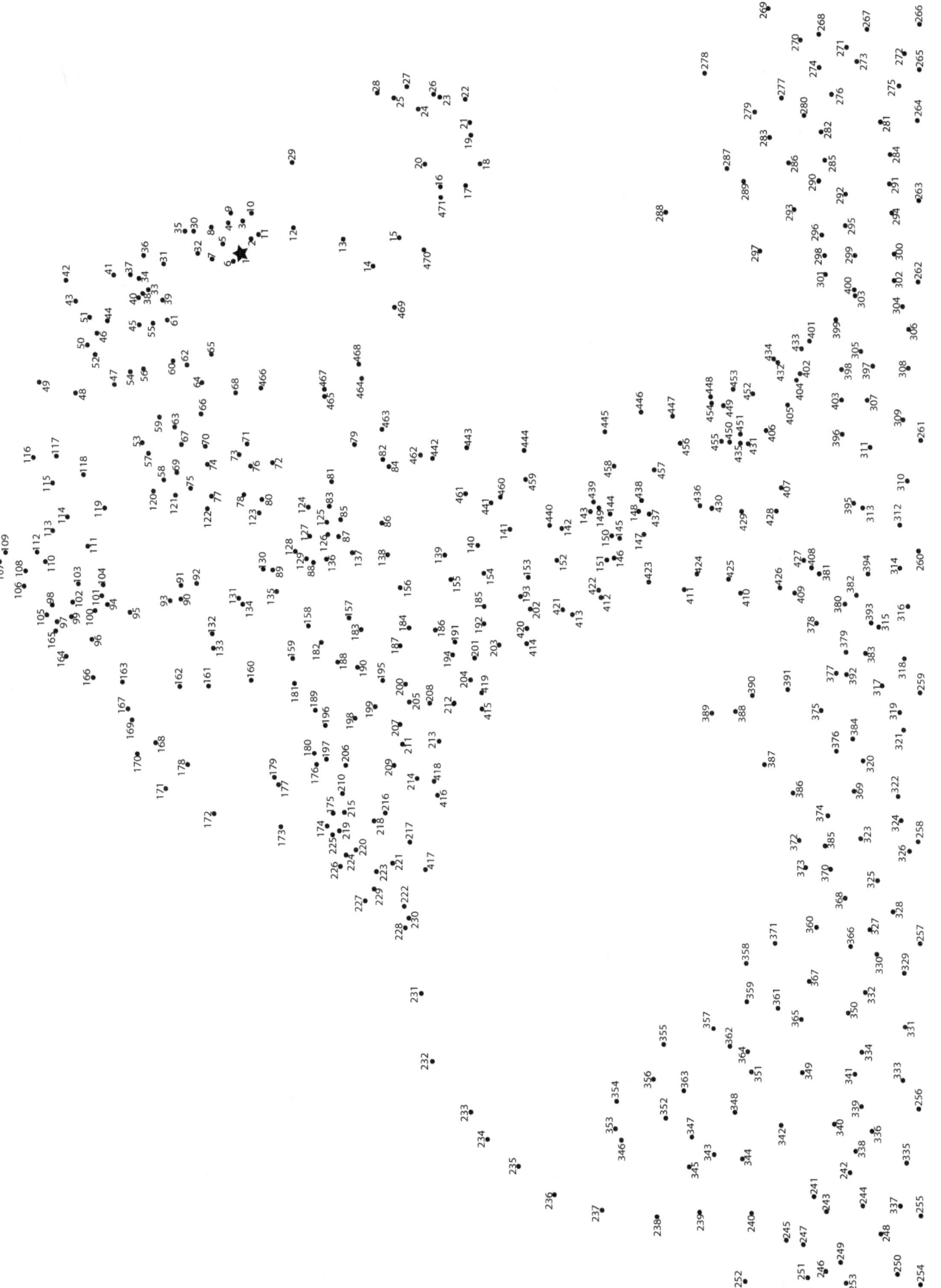

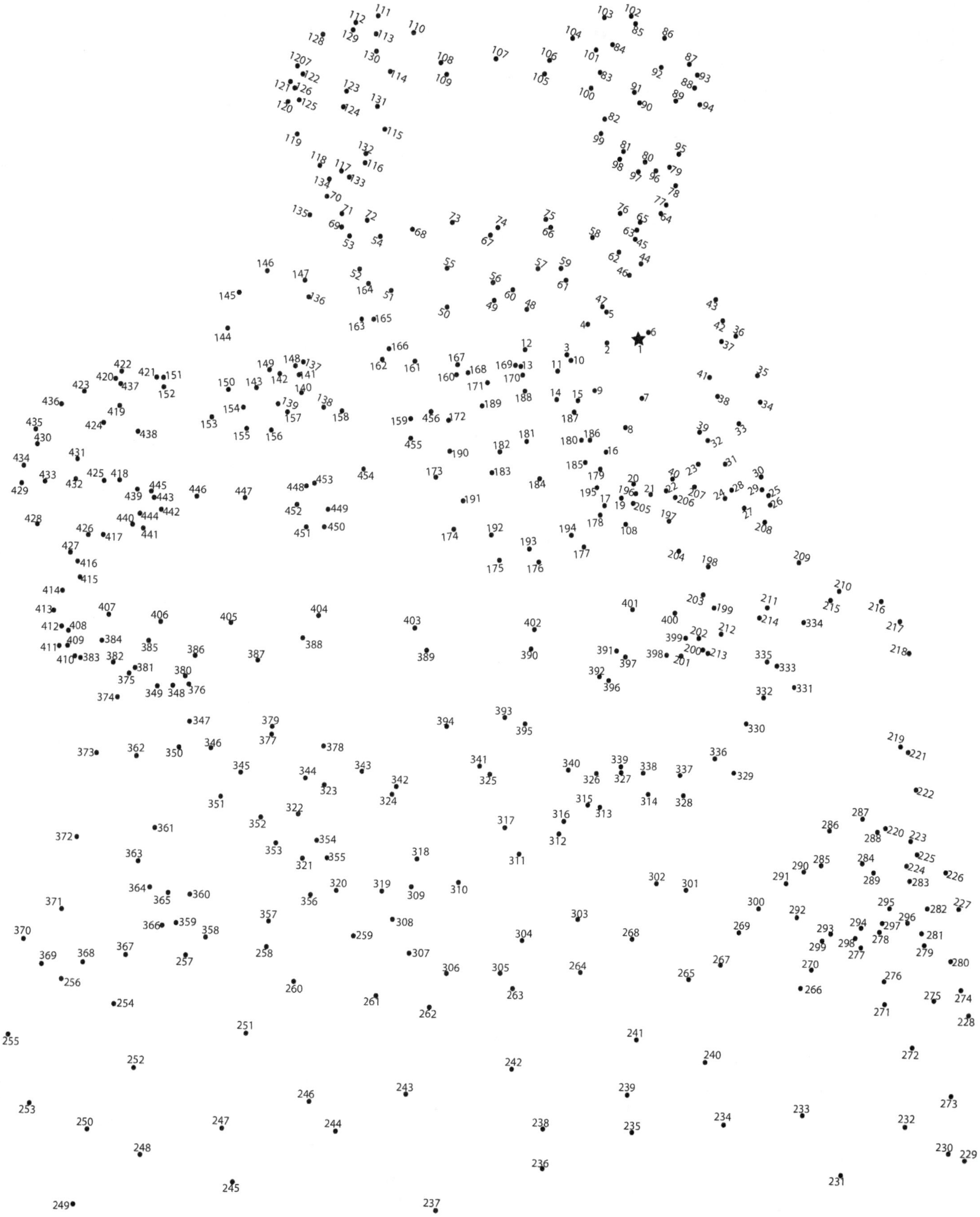

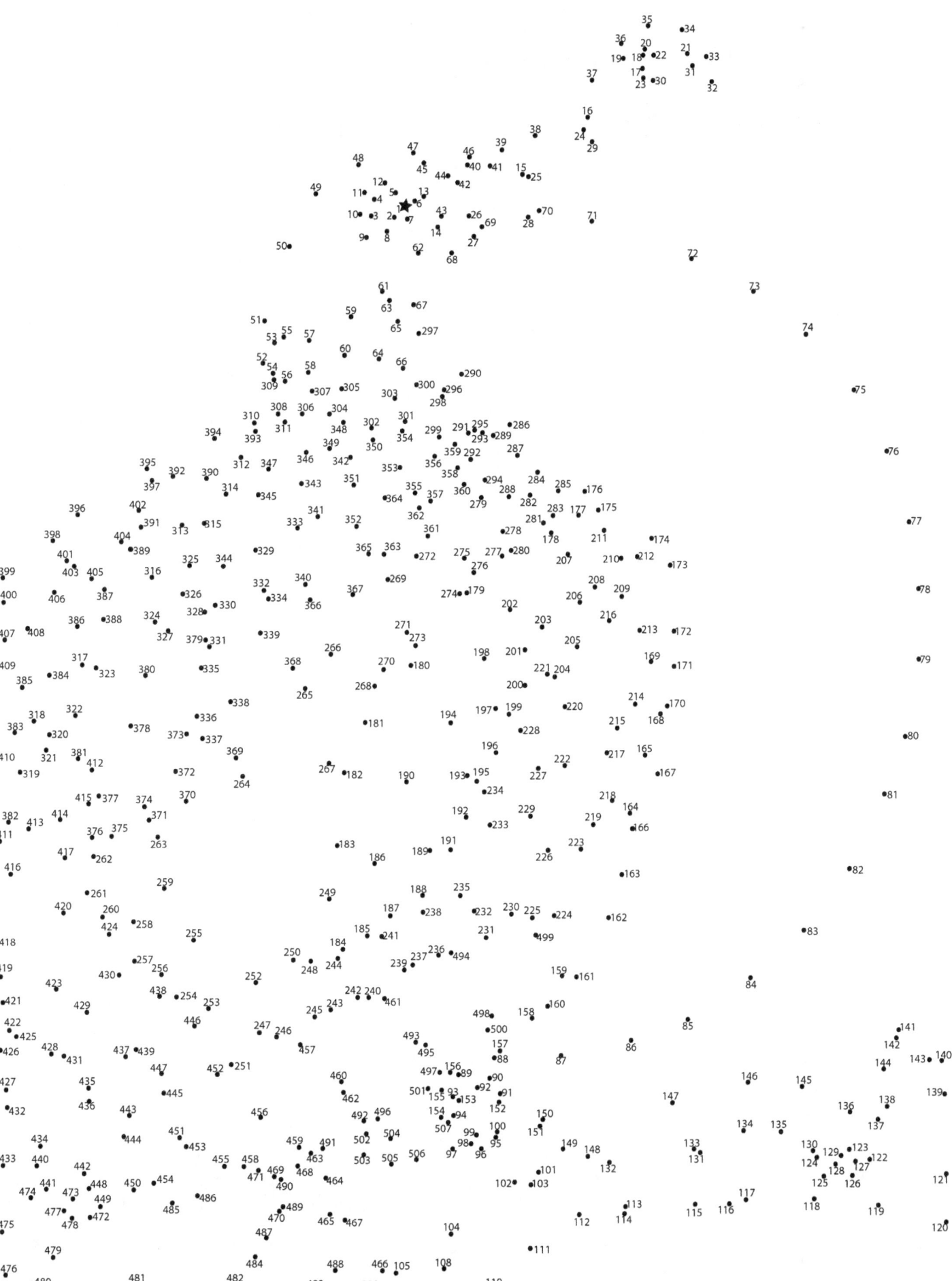

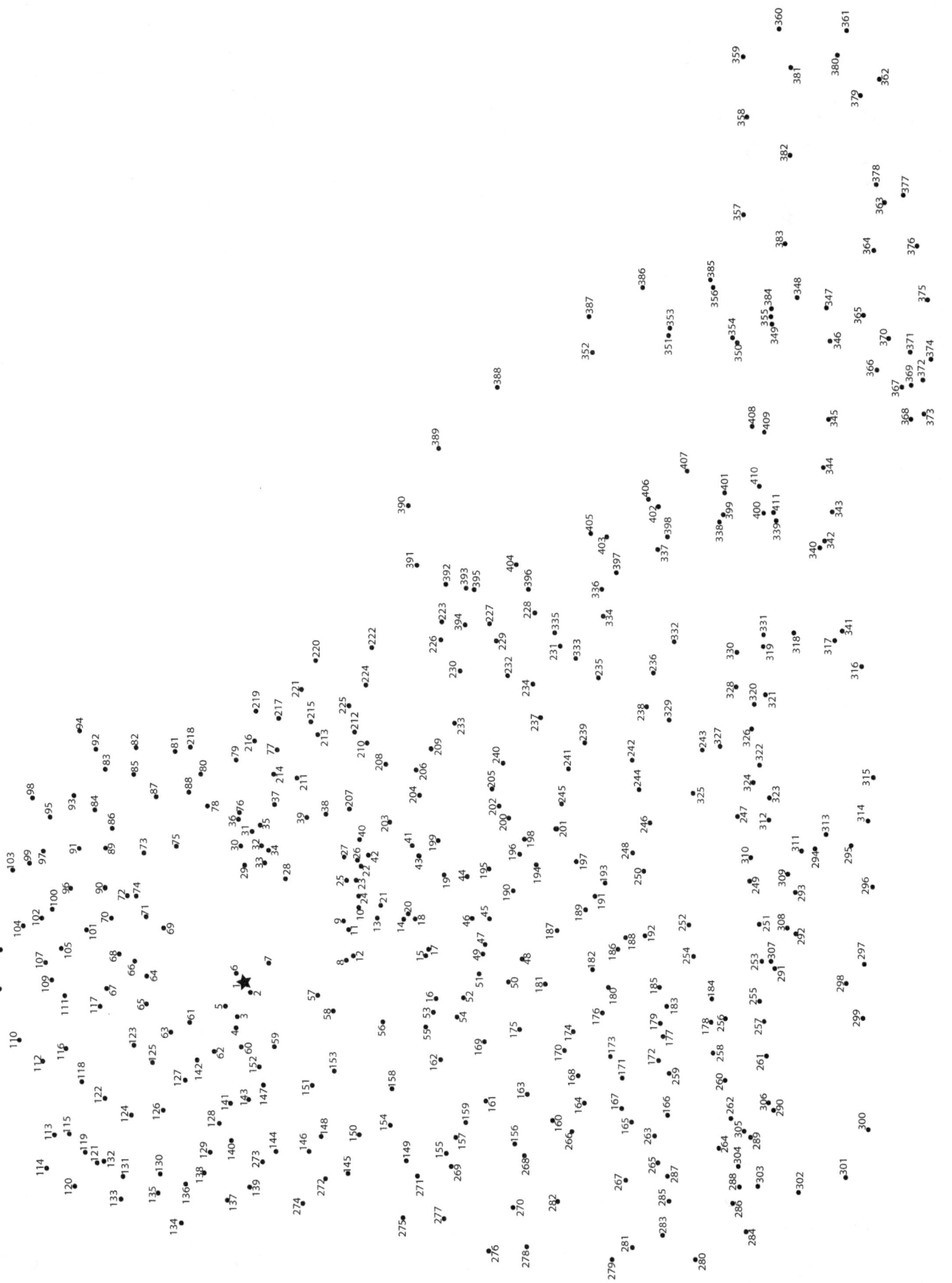

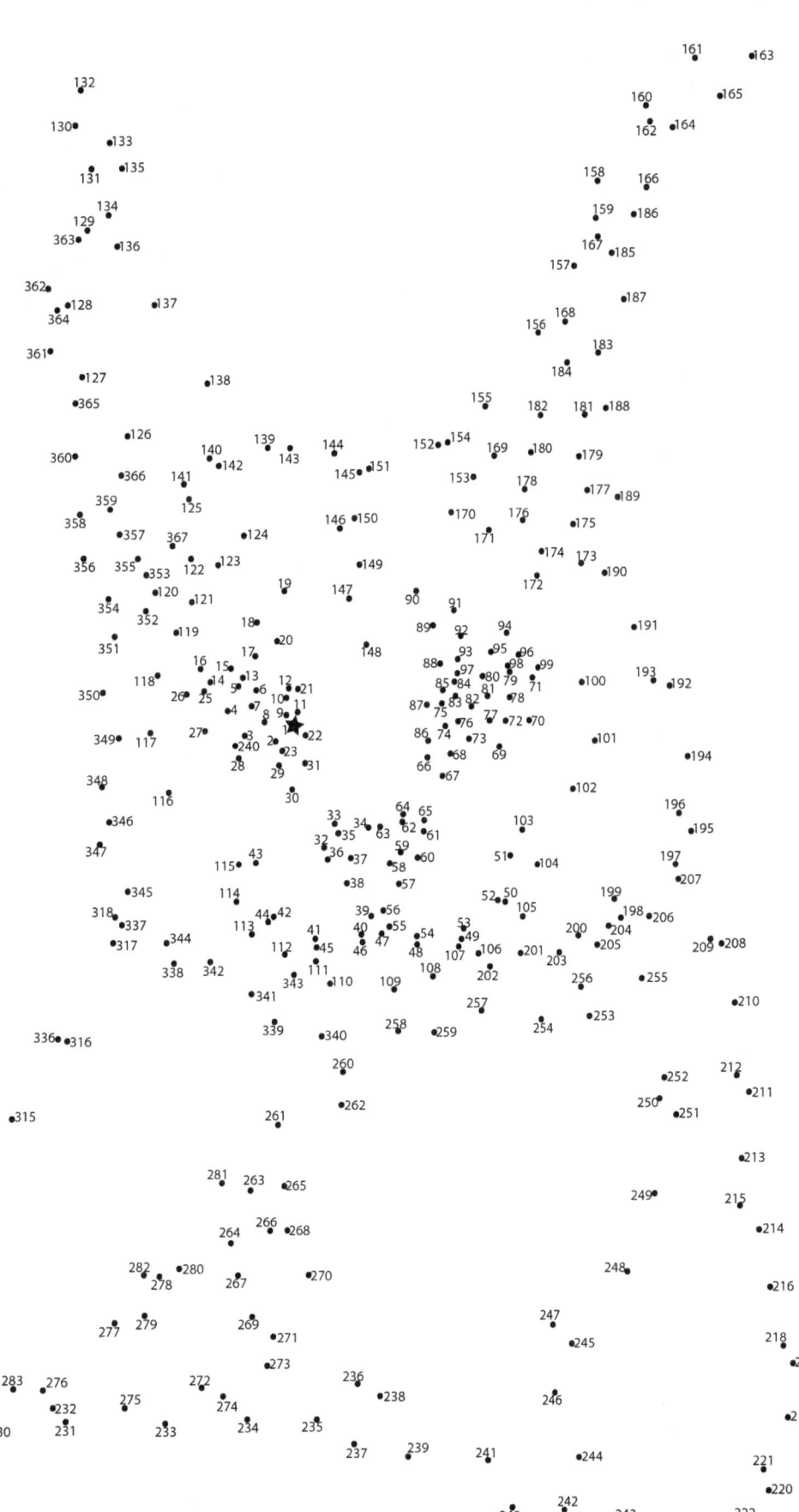

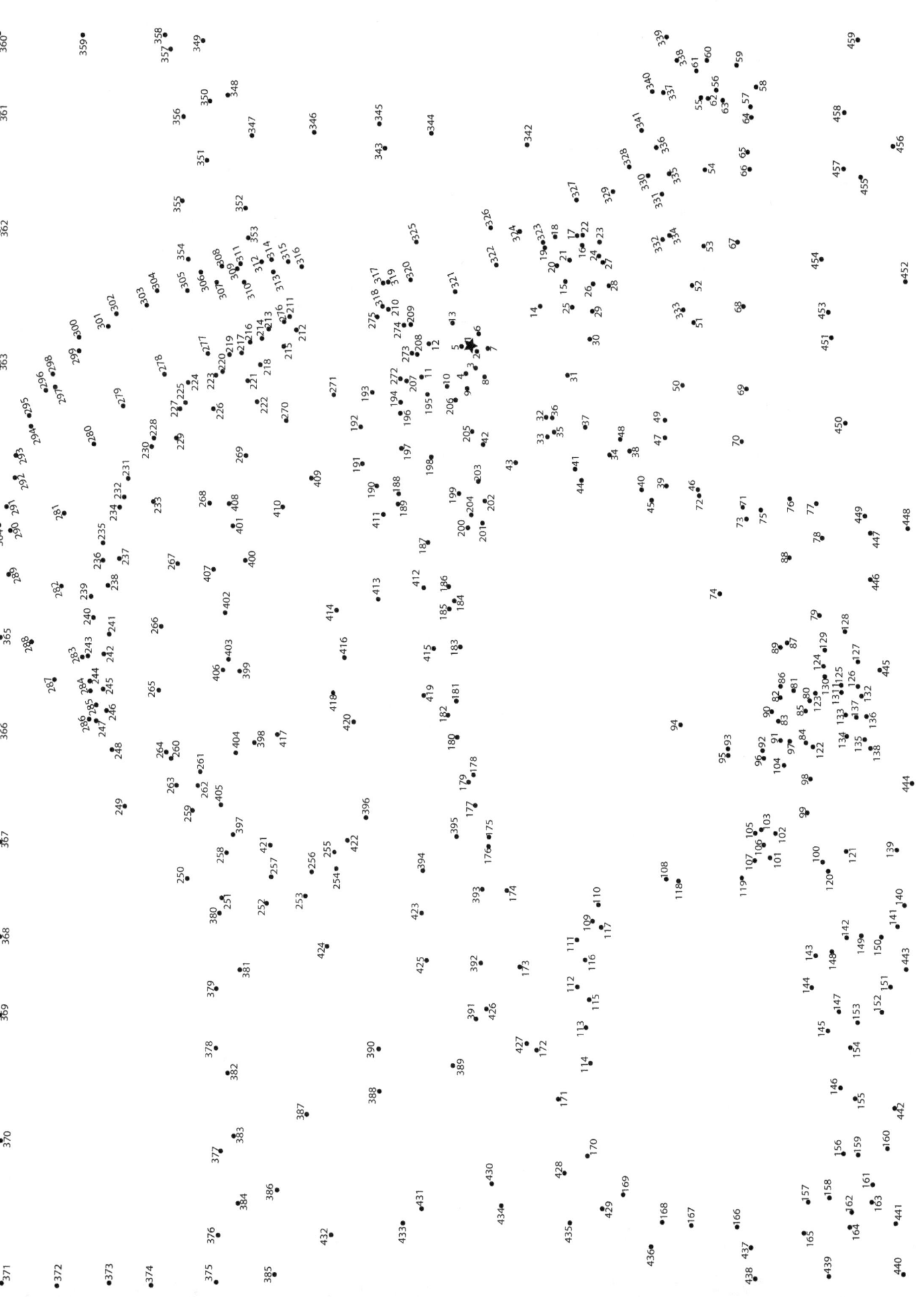

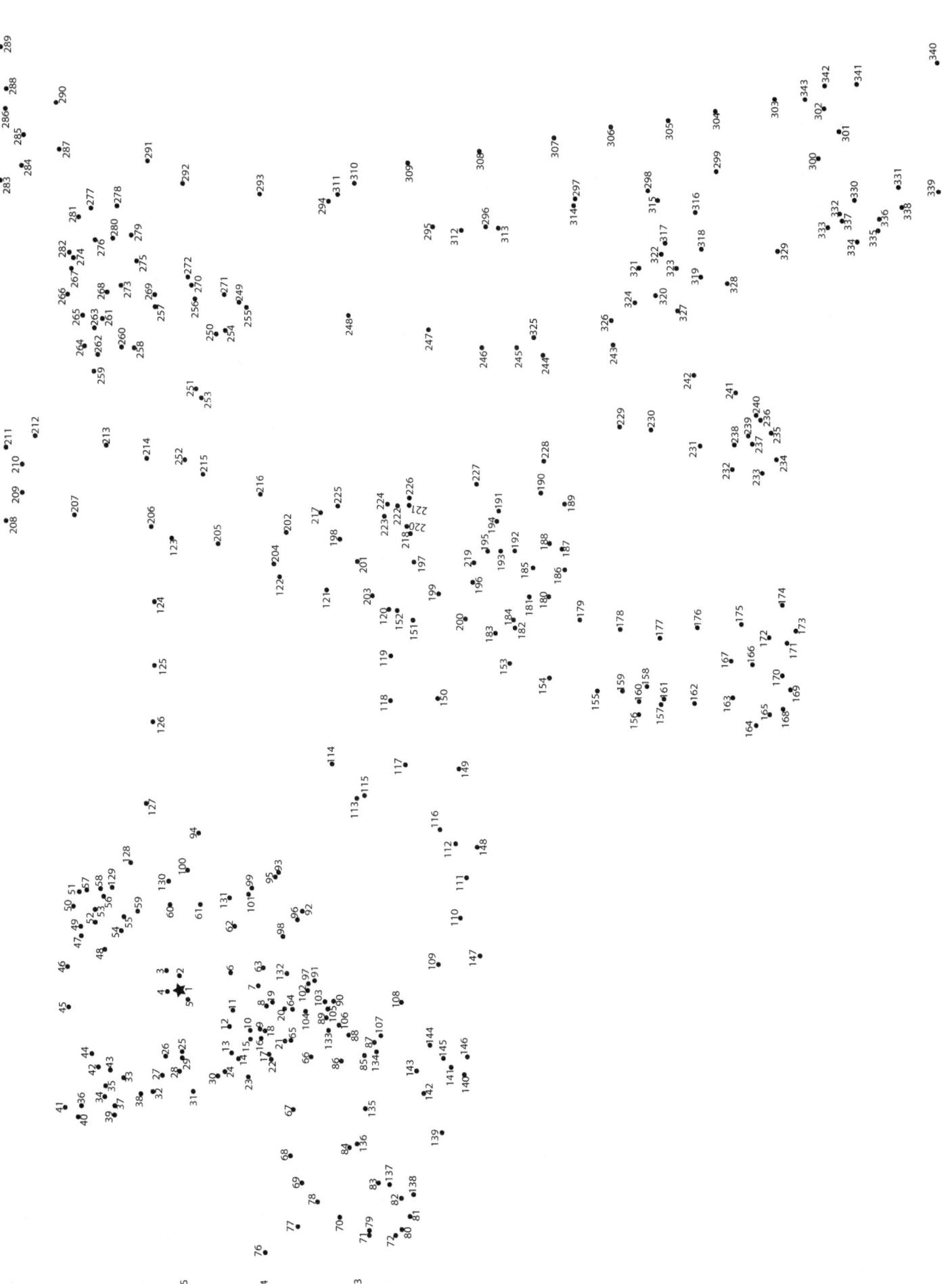

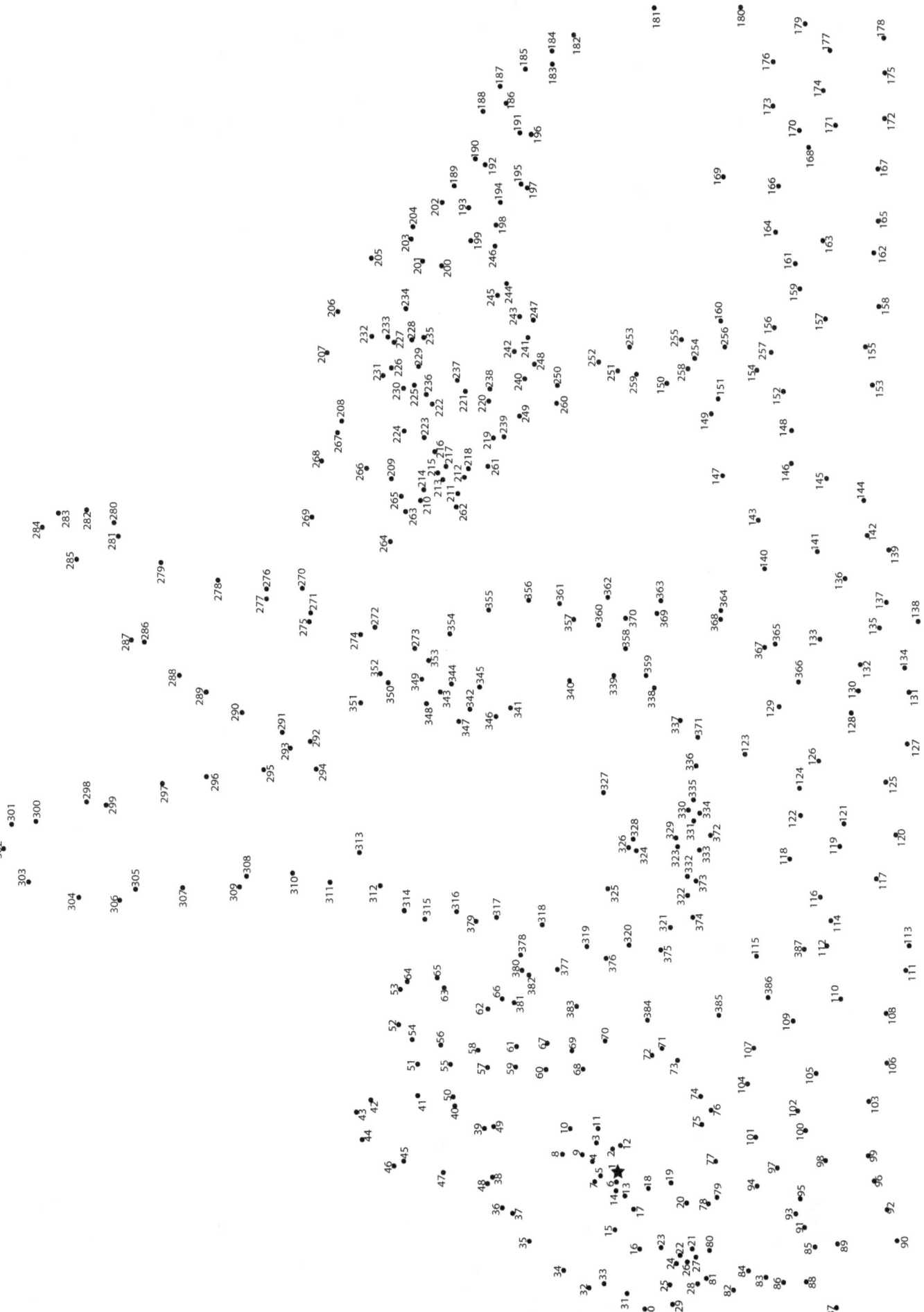

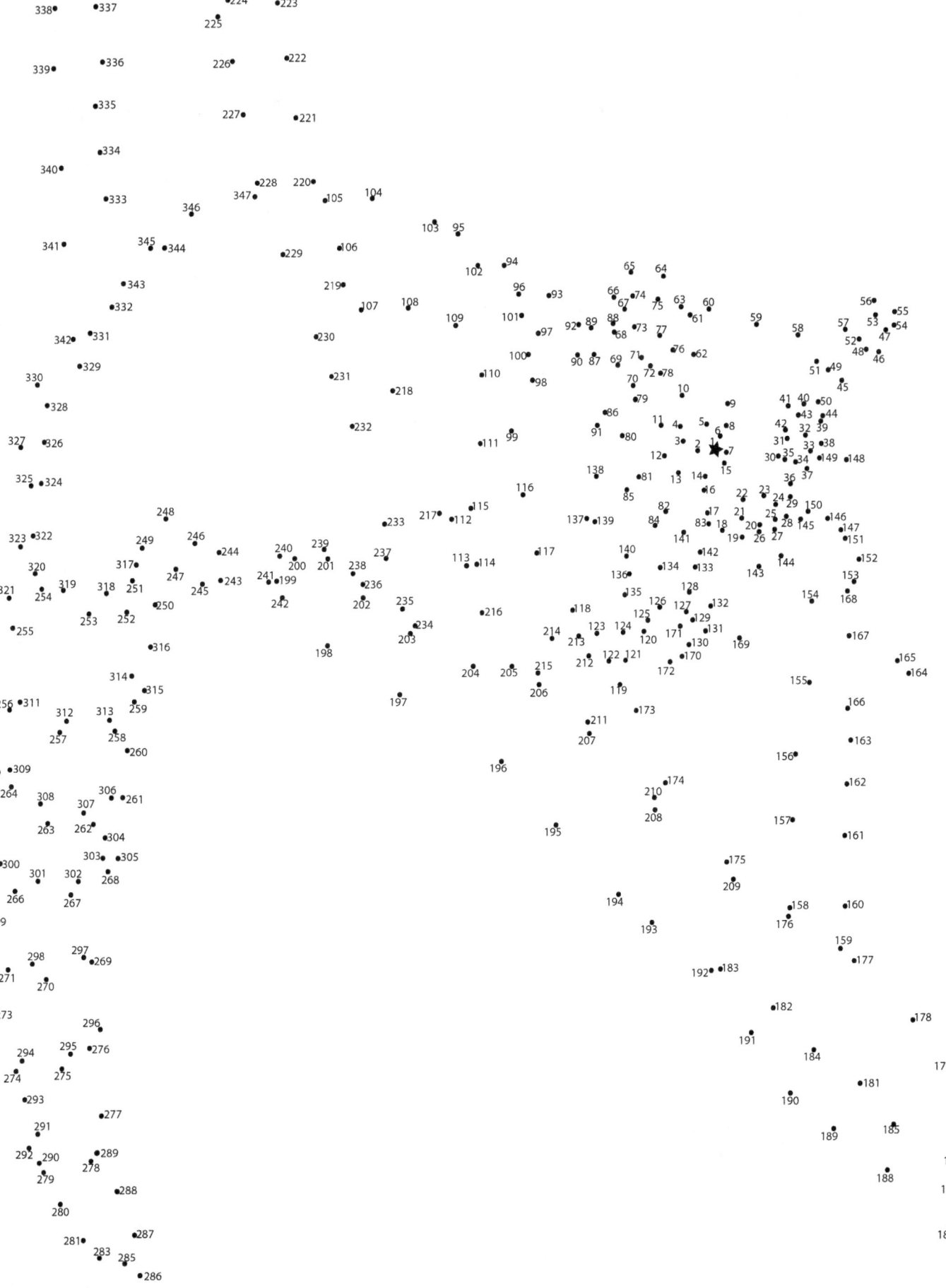

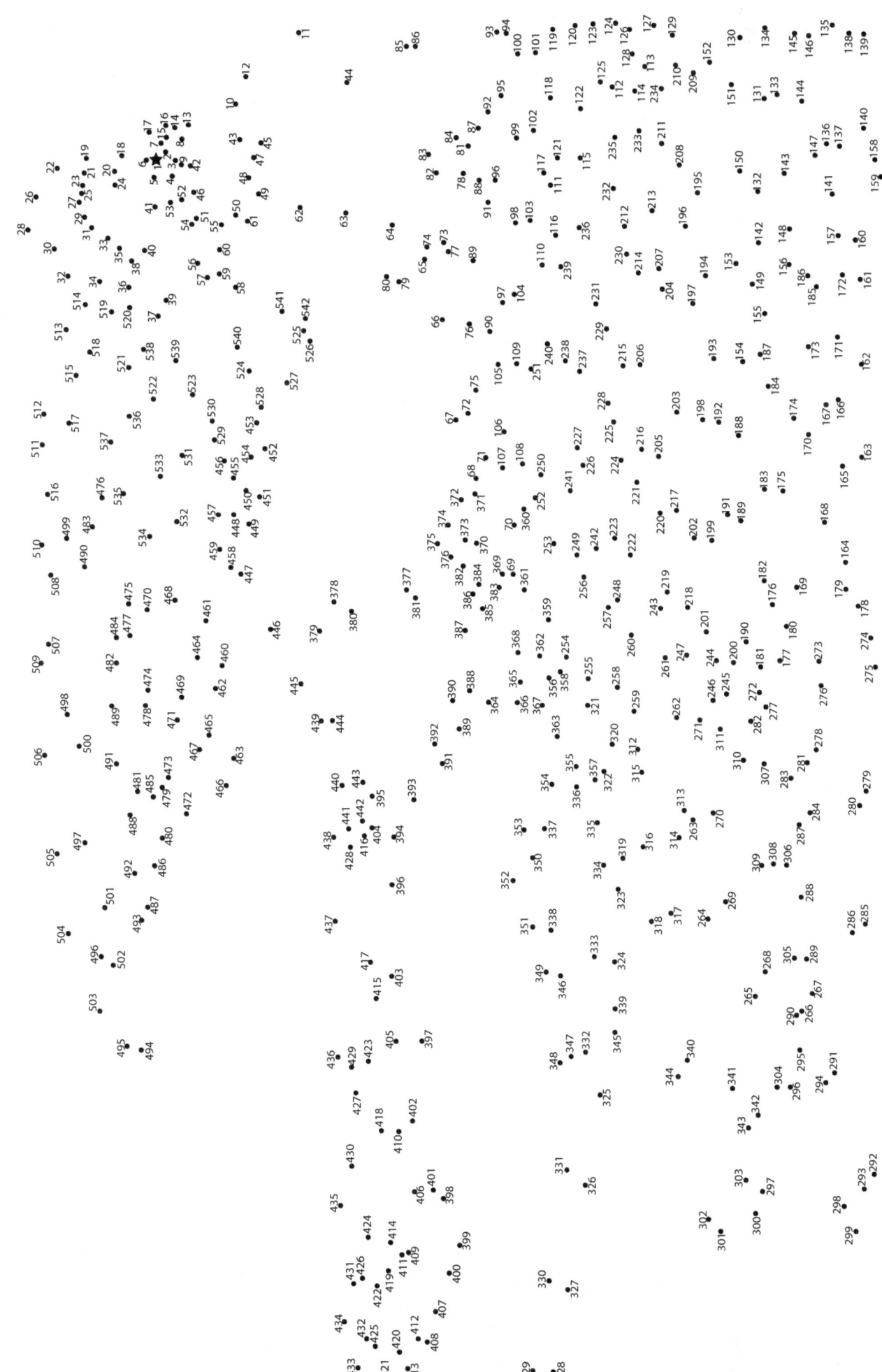

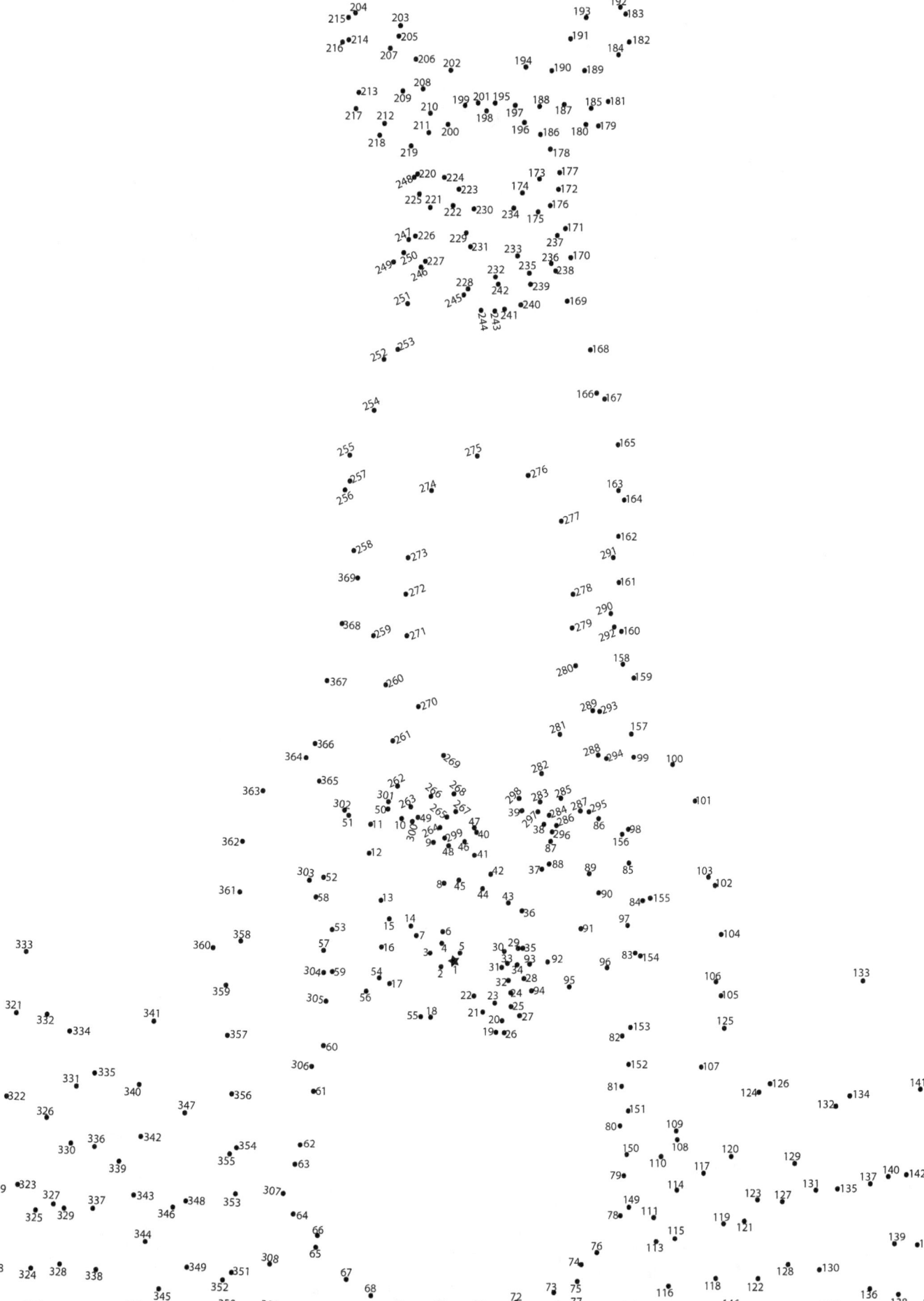

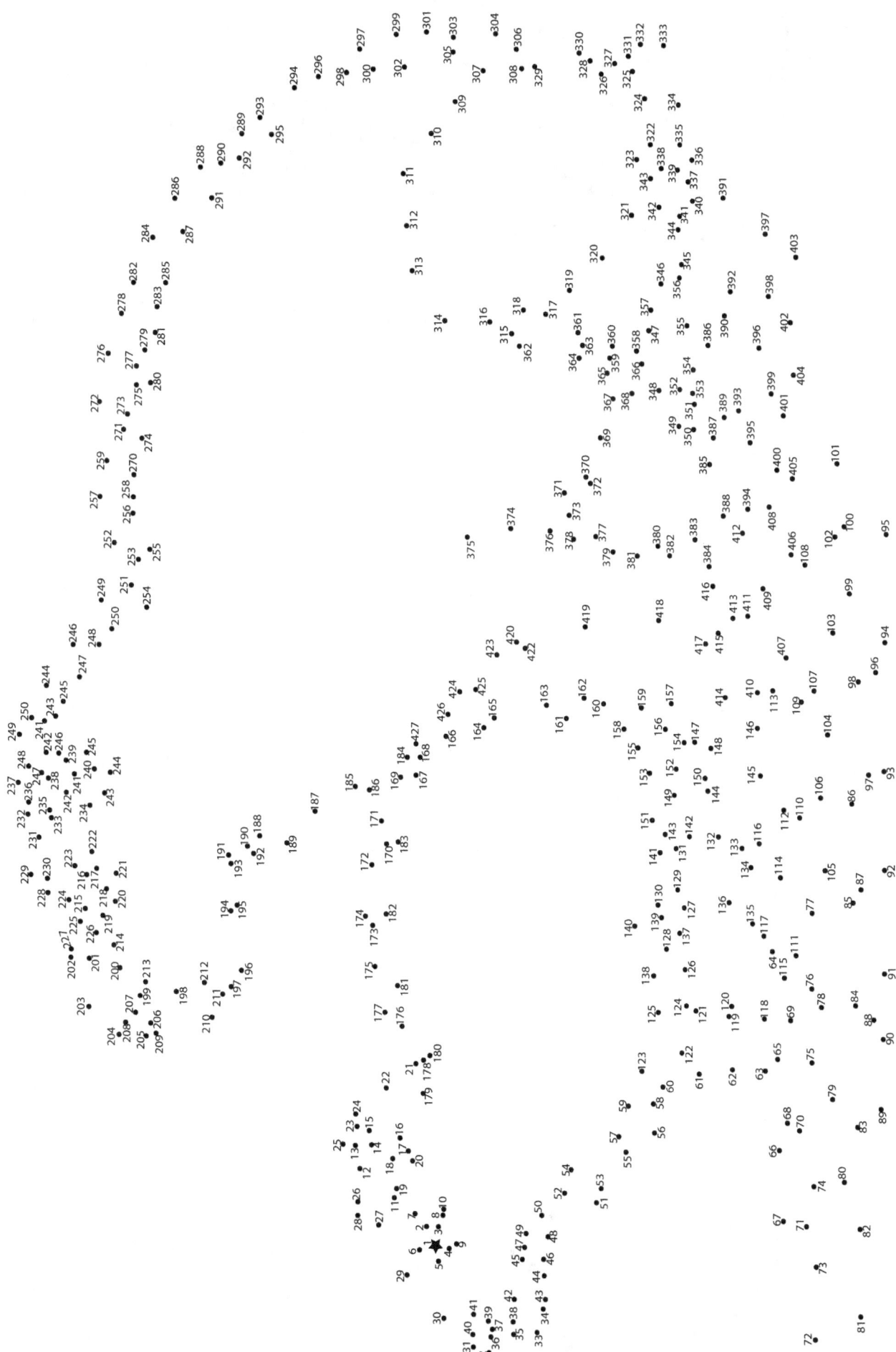

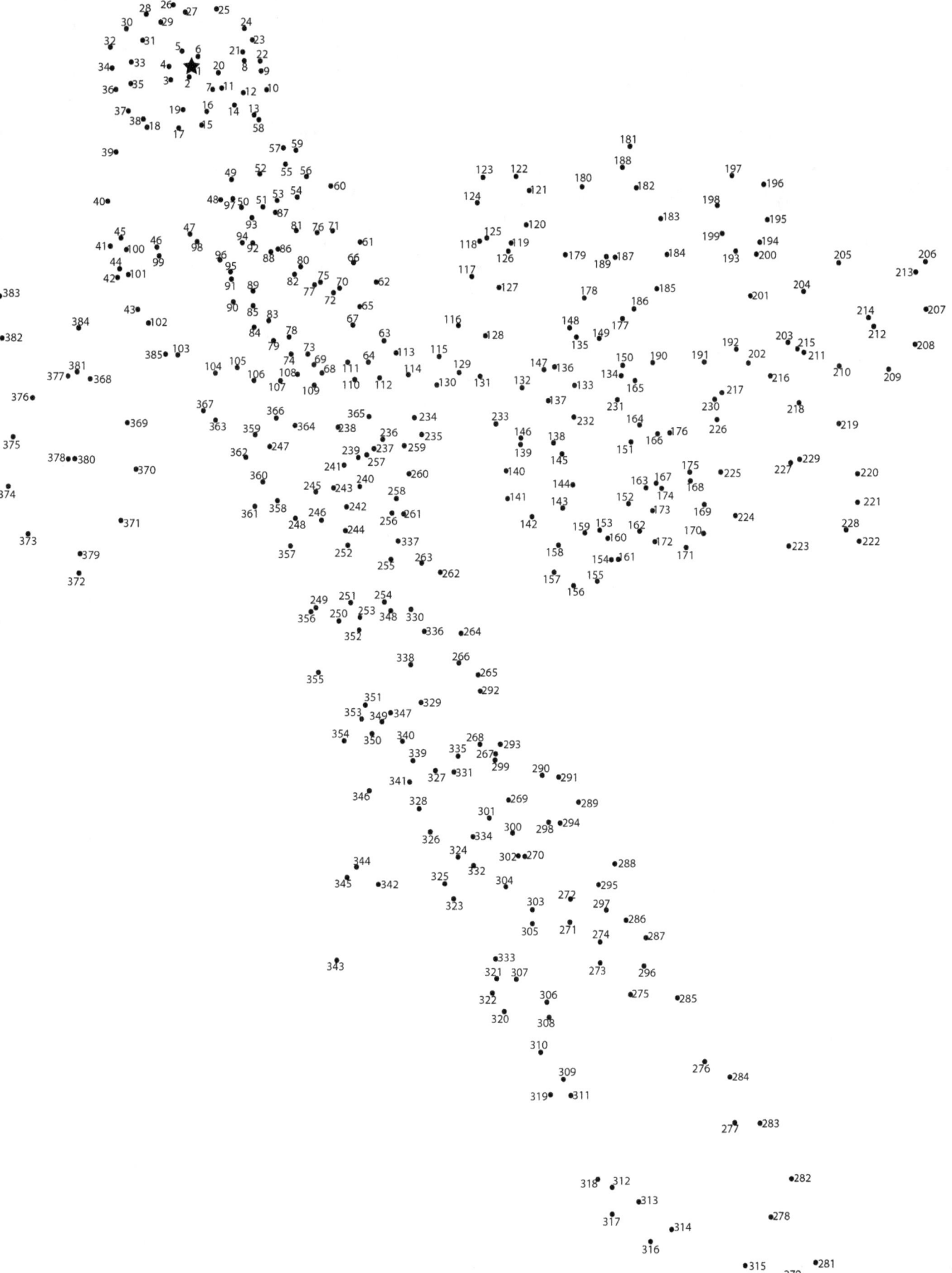

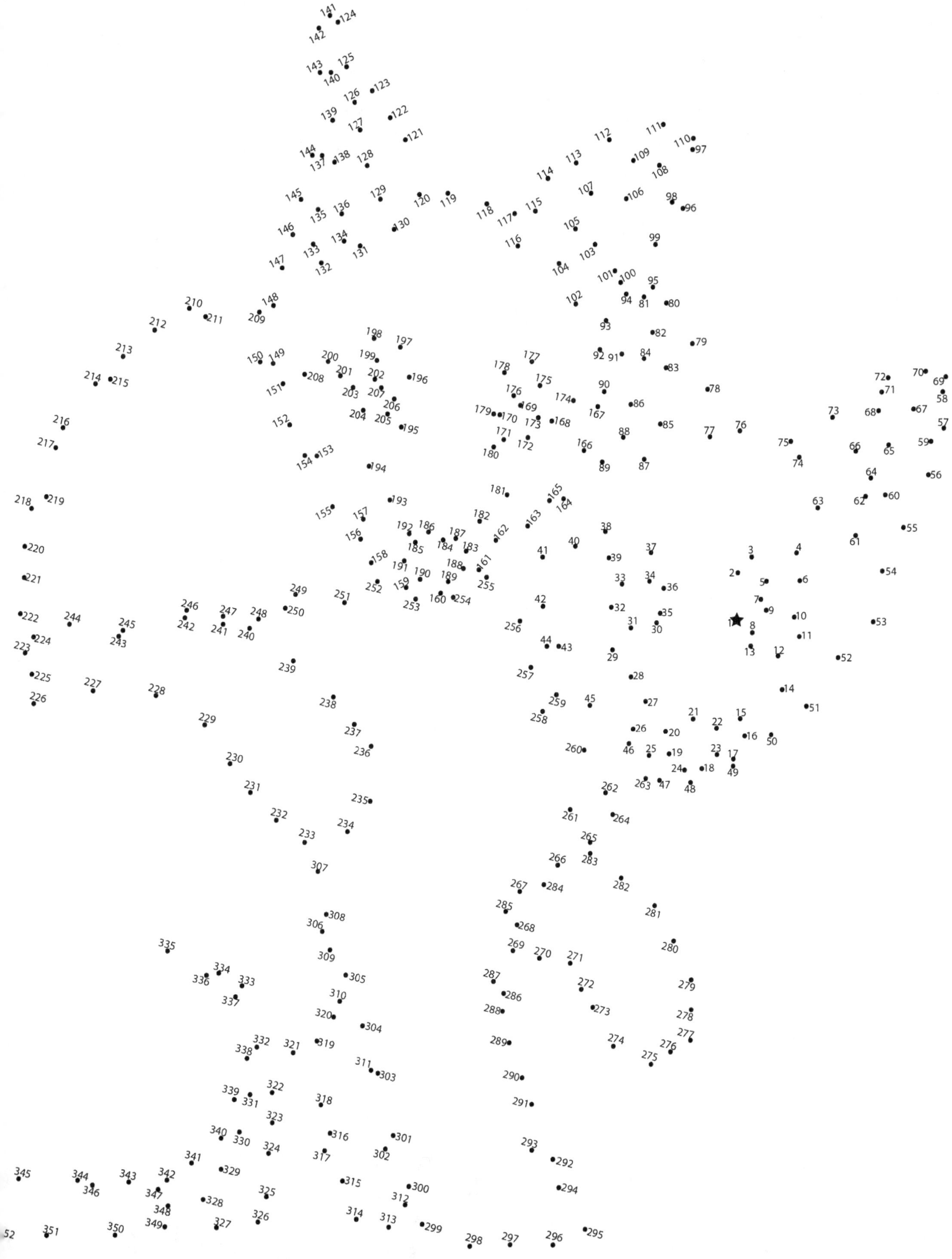

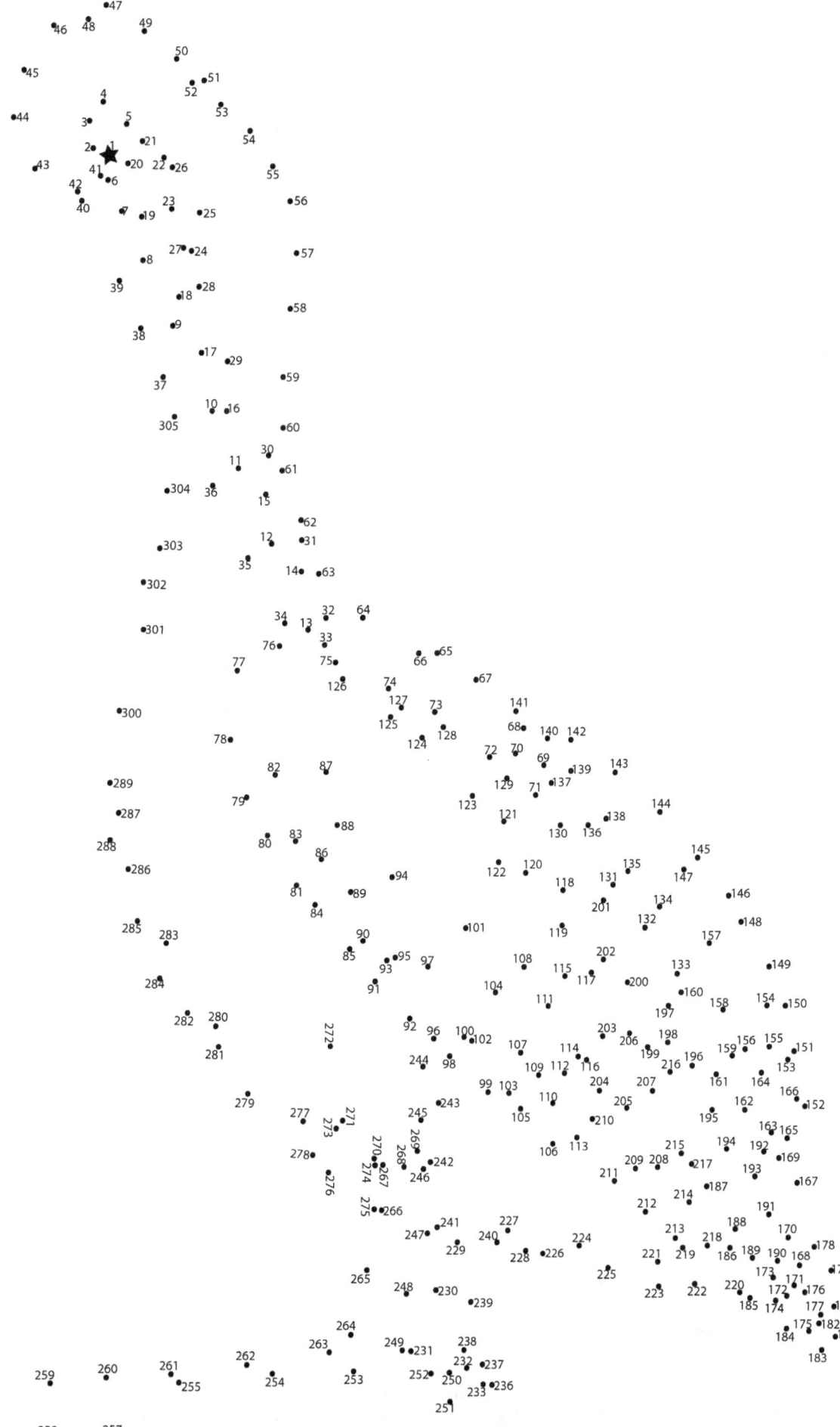

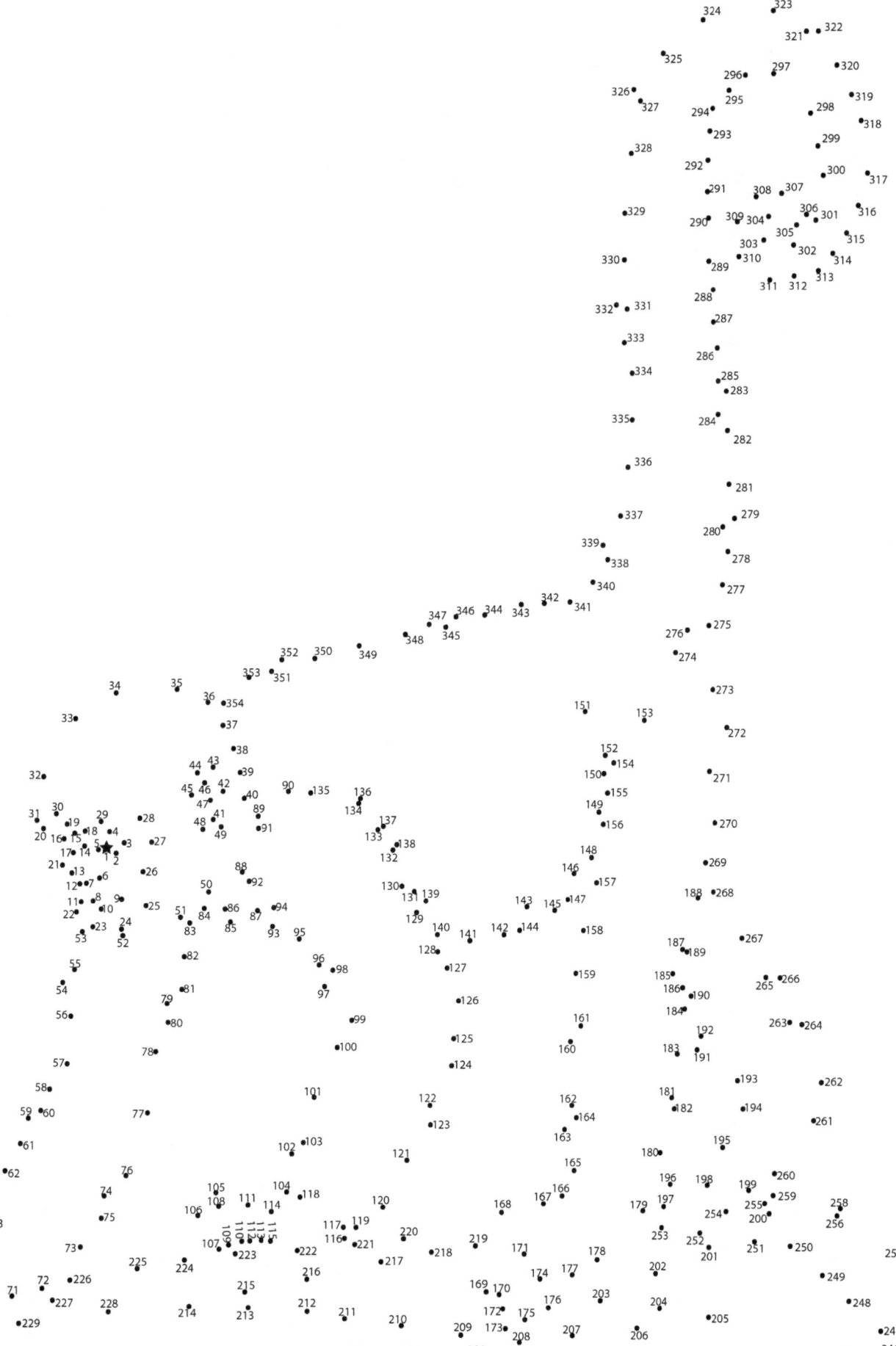

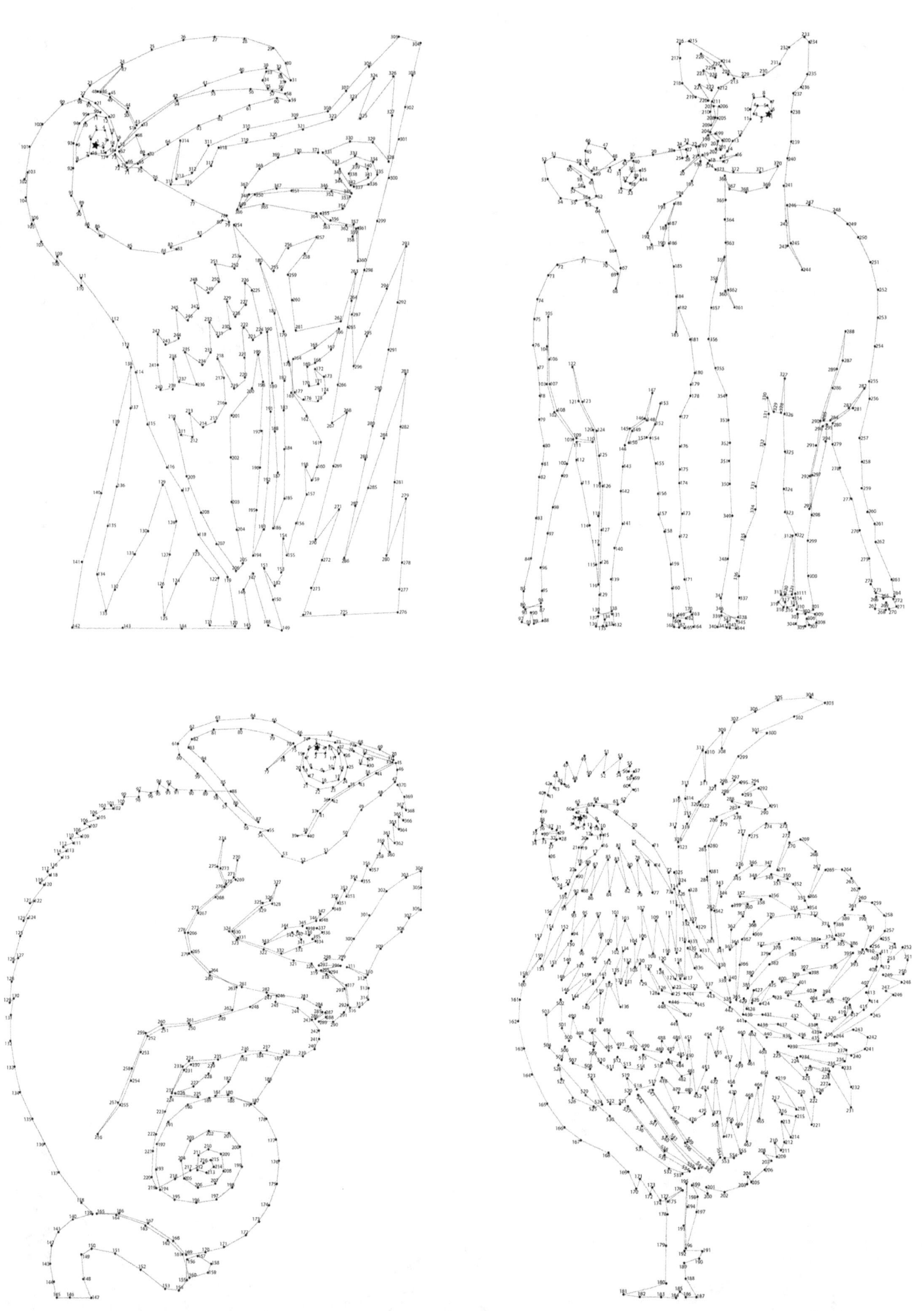

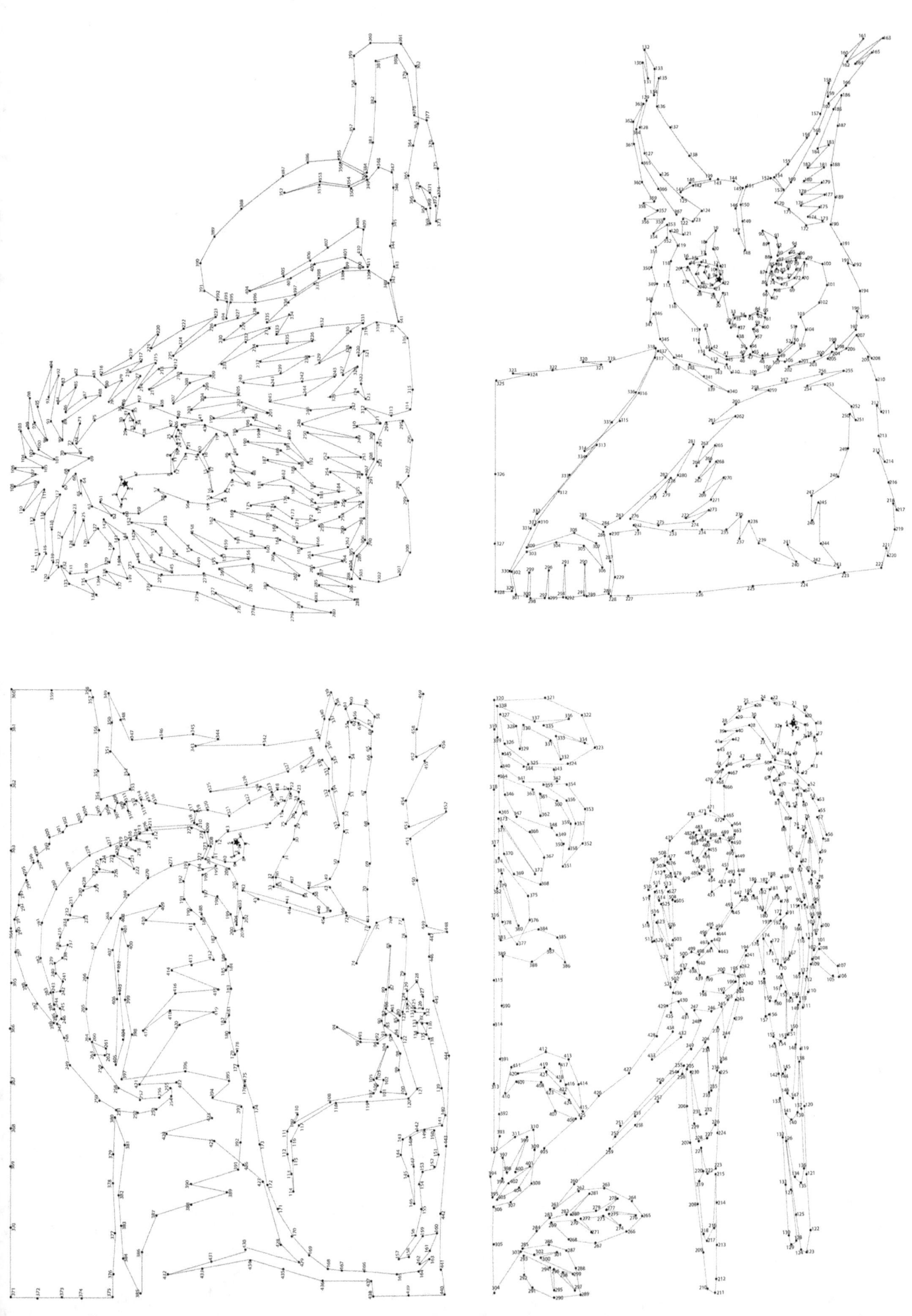

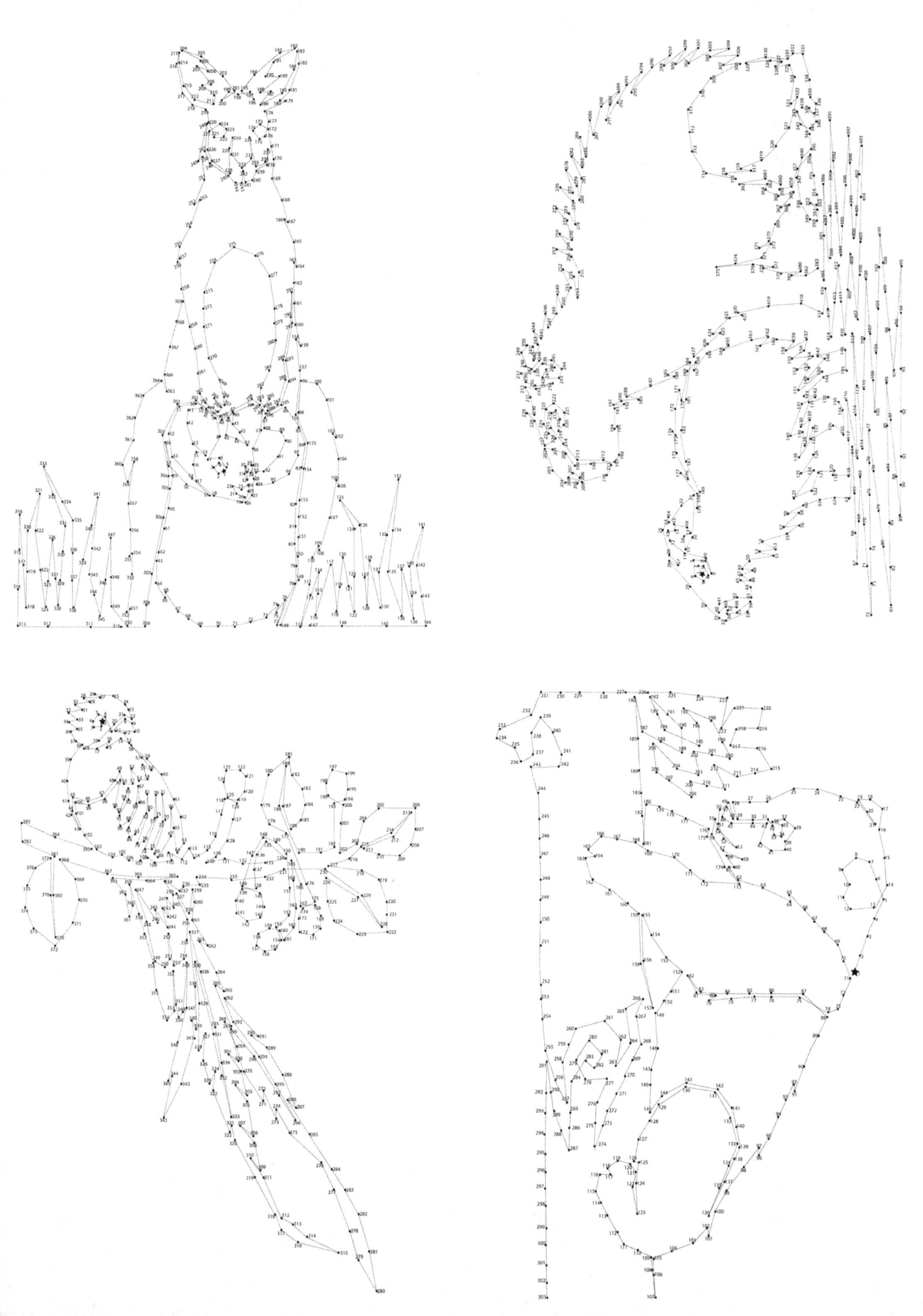

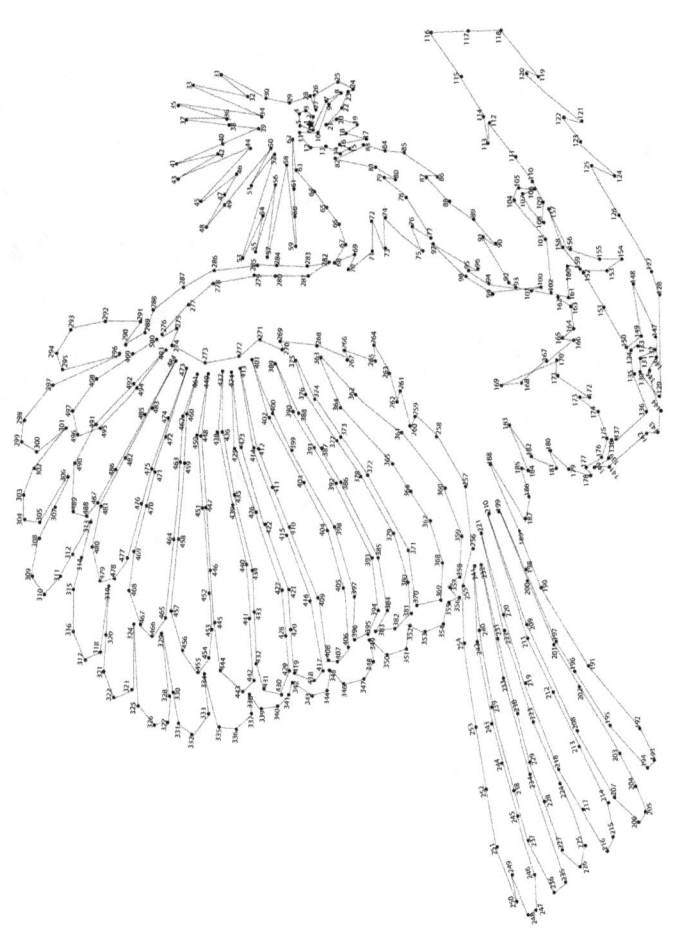

Enjoy these preview pages from some of our other books!

Connect The Dots For Adults
Birds Around The World

Dot to Dot Book for Adults
Butterflies and Flowers

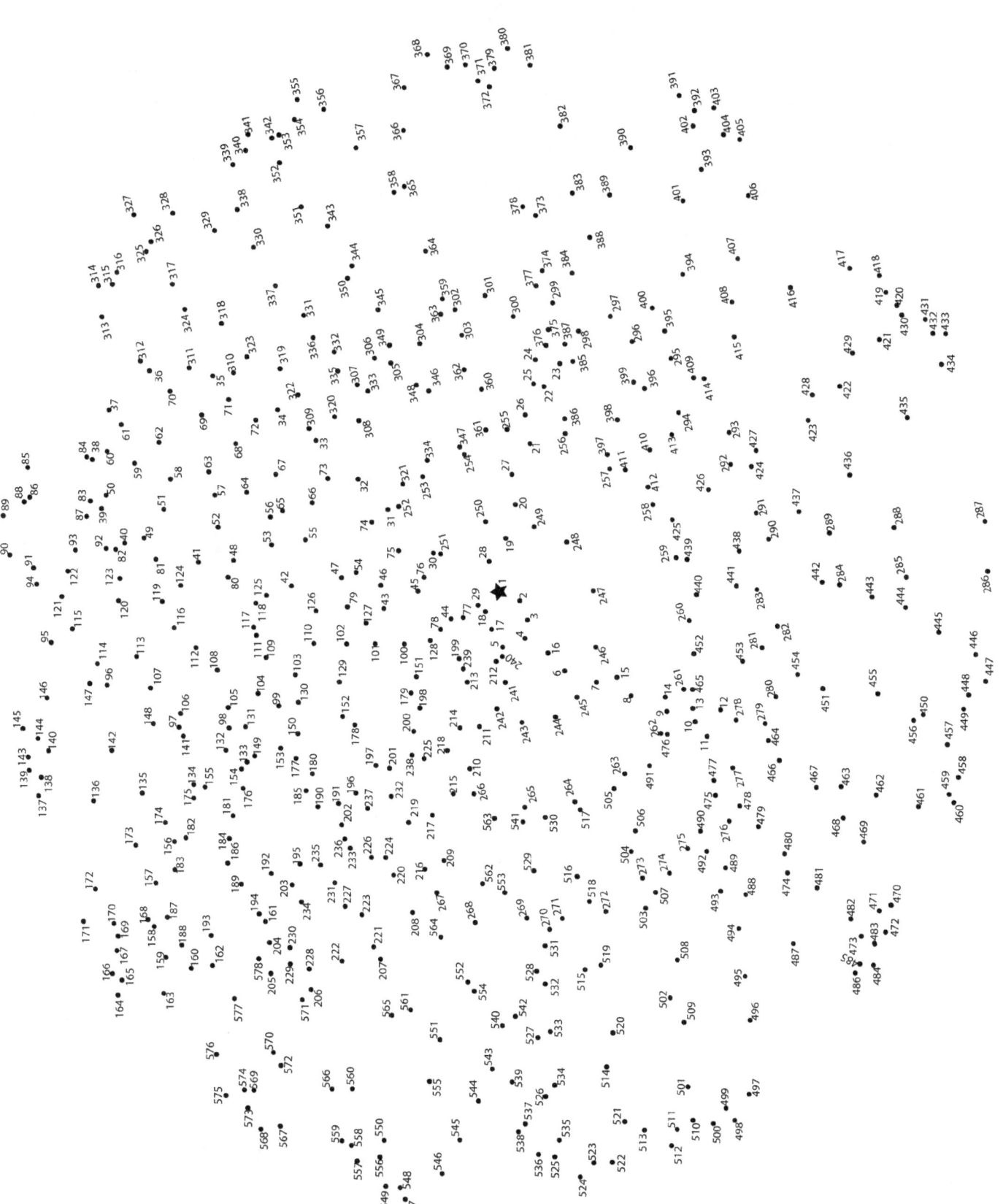

www.ingramcontent.com/pod-product-compliance
Lightning Source LLC
Chambersburg PA
CBHW081750220526

45468CB00008B/2315